GW00496871

Cover illustration: A Gefreiter of a Mountain Troop regiment in 1940, wearing Model 1936 tunic, snow goggles and Bergmütze. The weapon is a Dreyse MG13 light machine-gun.

1. Panzer-Grenadiers fighting in Hungary during the last months of the war. The need for camouflage was not as critical here as earlier on the Normandy invasion front.

UNIFORMS ILLUSTRATED NO. 5

German
Combat Uniforms
of World War Two
Volume I BRIAN L. DAVIS

ARMS AND ARMOUR PRESS
London—Melbourne—Harrisburg, Pa.—Cape Town

Introduction

Uniforms Illustrated 5: German Combat Uniforms of World War Two, Volume I
Published in 1984 by Arms and Armour Press, Lionel Leventhal Limited, 2–6 Hampstead High Street, London NW3 1QQ; 11 Munro Street, Port Melbourne 3207, Australia; Sanso Centre, 8 Adderley Street, P.O. Box 94, Cape Town 8000; Cameron and Kelker Streets, P.O. Box 1831, Harrisburg, Pennsylvania 17105, USA

British Library Cataloguing in Publication Data
Davis, Brian L.
German combat uniforms of World War Two. – (Uniforms illustrated; 5)
Vol. 1
1. Germany, *Wehrmacht* – Uniforms – Pictorial works
I. Title   II. Series
355.1′4′0943   UC485.G3
ISBN 0-85368-667-X

Editing and layout by Roger Chesneau.
Typeset by CCC, printed and bound in Great Britain by William Clowes Limited, Beccles and London.

Troops serving in the field, especially those under combat conditions, soon take on a distinctive, scruffy appearance – a far cry from the smart parade or walking-out uniforms worn by these same men at home or during periods of martial celebration. Although the German military authorities made every effort to try to provide their troops with warm, dry quarters and with facilities to keep clean and shaven and their clothing and equipment in a state of good order and repair, it was almost impossible to maintain these requirements throughout the armed forces under wartime conditions. As the war progressed the state of uniforms became worse. The availability of replacement garments became more difficult and extemporized uniform items more commonplace. Something of this problem can be gauged from photographs 2 and 3: the first was taken in 1942, when it was still possible for German personnel to appear smart under active service conditions; the second was taken in the summer of 1944, and shows some of the first prisoners to have been captured at Calais arriving in the United Kingdom.

The photographs featured in this book should give some idea of the appearance of regulation clothing and uniforms worn by German troops in many theatres of war, at different seasons of the year, and at different periods during the hostilities.

Brian L. Davis

◄2
2. Hauptmann Hirschfeld receiving the award of the Iron Cross 1st Class from the hand of the Commander-in-Chief West, Generalfeldmarschall von Rundstedt, 26 August 1942. Captain Hirschfeld received his award for the part he played in countering the British and Canadian landings at Dieppe.

◀3

4▲

3. Two second-line German prisoners-of-war arriving at a Channel port on their way to a PoW cage somewhere in England. Unkempt, and wearing shoddy and patched uniforms, these men were not untypical of the type of prisoner that was being captured in ever greater numbers during the last months of the war in Europe. The very poor quality of the uniform material is evident in this photograph. The M44-pattern blouse being worn by the soldier on the left was introduced to economise on material and labour costs, and it was quite a departure in design from the four-pocket Field Service tunic worn before 1944. Both men wear ankle boots, an economy brought about by the need to save on the leather used in the manufacture of marching boots (the familiar 'jack boots'), and even their caps are a wartime introduction.

4. Aggressive in attack, tenacious in defence, the German Infantry bore the main burden of all fighting. Here, German combat troops cautiously wade into shallow waters at a crossing point on the River Don.

5. Infantry assault troops move forward during the fighting for Charkov.

5▼

▲6 ▼7

6. Infantry on the move, taking part in a defensive battle somewhere on the Eastern Front in August 1943.
7. Troops pause for a rest during a long march, May 1942.
8. Assault Infantry crossing the River Don by power-driven inflatable rubber dinghies.
9. German Grenadiers anxiously scan the skies for signs of Allied aircraft during a pause in the fighting for Normandy in the summer of 1944.

8▲ 9▼

10. A detachment leader from a heavy machine-gun company observing enemy positions through camouflaged scissors binoculars. His winter uniform is worn with the mouse-grey side outermost, and he is also wearing padded leather gloves and the Officer's Old Style Field Service cap.

11. Linesmen from a German DAK Signals Unit laying down field telephone wires somewhere in the North African desert.

12. Invasion front, Normandy: Grenadiers confer with a tank commander.

◀10   11▶

12▶

▲13 ▼14

13. An Infantry foot patrol of seemingly very young troops moving across an open field somewhere in the Normandy area.

14. An August 1943 photograph showing an unusual combination of uniform items: a sentry stands guard on a beach somewhere on the southern part of the Atlantic Wall wearing the standard-pattern temperate climate Field Service uniform with a tropical sun helmet.

15. German Infantry enter a Belgian town in May 1940. Whilst these troops have the appearance of soldiers who have been serving under active service conditions, they look positively smart when compared to the pictures of German soldiers taken a few years later in the war, especially on the Eastern Front.

16. An Oberfeldwebel demonstrating the newly introduced 'Panzerfaust', the German answer to the American 'Bazooka' anti-tank, hand-held weapon and the British 'PIAT'. Worn on the right forearm of the Field Service tunic is the German Army trade badge for a 'Feuerwerker', a yellow gothic letter 'F' on a dark blue-green circular patch edged in silver cording. A Feuerwerker was an NCO Artificer or Ordnance Technician.

◀16

▲17 ▼18

17. General-feldmarschall von Brauchitsch, the Commander-in-Chief of the German Army from 4 February 1938 to 19 December 1941, inspects German Mountain Troop personnel in Greece. As well as wearing his Knights Cross award at his neck and the Iron Cross, 1st Class, pinned to his left breast pocket, he also wears the Gold Party Badge. This Nazi Party decoration was presented to those members of the NSDAP who bore a membership number lower than 100,000. Other awards of this badge were made to persons considered to be deserving of such an honour even though they had joined the Nazi Party long after the first hundred thousand members.

18. A camouflaged anti-tank gun emplacement somewhere on the Charkov front. The sentry is wearing an unusual garment which appears to be a heavy-duty waterproof with combined hood.

19. German Infantry prepare to go on patrol. The weather is obviously cold, and light snow is still on the ground. The men are wearing greatcoats for warmth, over which are white 'snowshirts' as a simple form of snow camouflage. The crude white cloth covers worn over the steel helmets add to the camouflage effect.

20. Making good use of the cover afforded by a destroyed Red Army Sherman, German tank and artillery observers study the ground ahead held by the Russians: the summer of 1944, when this photograph was taken was warm enough for troops to be lightly clad (and here, in one instance, to be in shirtsleeve order). The man in the centre of the group of three is wearing the grey shirt of armoured personnel.

21. Troops making good use of a horse-drawn cart. The cart is loaded with all their kit and they are left free to walk behind without having to hump around the items that it was normally their task to carry. It is of interest to observe that, despite their good fortune to have this means of transport to hand, they are still cautious enough to carry their gas mask canisters.

22. All items illustrated in these colour plates were worn on the German Army Field Service uniform by troops of different rank and of different units, and at various times. Shoulder straps and collar patches were worn in matching pairs, but for the sake of brevity only one strap or patch has been illustrated from each pair.

A. Cuff-title worn by troops serving in North Africa as the Deutsches Afrika Korps; worn on left forearm.

B. Shoulder strap for a Generalmajor.

C. Collar patch (left side) for an official of the Army High Command.

D. Collar patch (left side) for senior Army officers with the rank of Generalmajor, Generalleutnant, General der Infanterie etc., and Generaloberst.

E. Trade badge for a Signals Operator of an Engineer (Pioneer) Unit; worn on upper left sleeve.

F. Trade badge for a Signals Operator from an Infantry Unit; worn on upper left sleeve.

G. Trade badge for a Signals Operator from a Transport Supply Unit; worn on upper left sleeve.

H. Trade badge for a Signals Operator from a Motorised Reconnaisance Unit; worn on upper left sleeve.

I. Trade badge for a Radio Operator from a Cavalry Unit; worn on upper left sleeve.

J. Early-pattern shoulder strap for soldier of an Anti-Tank Unit.

K. Early-pattern shoulder strap worn by Other Ranks on the staff of Army Ordnance Technicians' School.

L. Duty gorget (Ringkragen) for troops of Military Field Police Units (Feldgendarmerie); worn suspended around the neck.

M. Arm badge for a qualified helmsman of engineer assault boats; worn on upper left sleeve.

N. Collar patch (left side) for an Engineer Official on the High Command.

O. Shoulder strap for a Veterinary Oberfeldwebel.

P. Cuff title for members of the Feldgendarmerie; worn on left forearm.

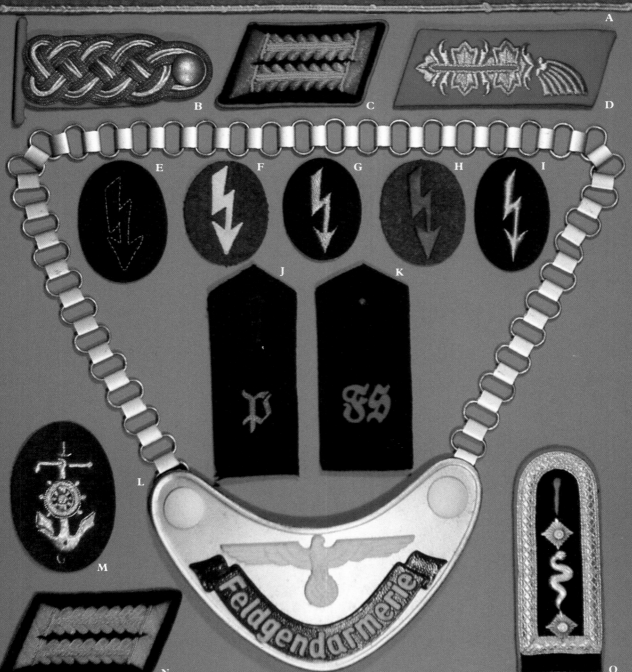

A

B

C

D

E

F

G

H

I

J

K

L

M

N

O

P

B

C

D

E

F

I

J

K

G

L

H

M

N

O

P

Q

R

Hilfs-
Krankenträger

S

23.
A. Cuff title for staff members of an Army NCO Preparatory School; worn on right forearm.
B. Rank badge for an Oberschütze; worn on left upper arm.
C. Rank chevron for a Gefreiter; worn on left upper arm.
D. Rank chevrons for an Obergefreiter with less than six years service; worn on left upper arm.
E. Rank chevrons for an Obergefreiter with more than six years service; worn on upper left arm.
F. Rank chevrons for a Stabs Gefreiter; worn on upper left arm.
Items B to F inclusive were rank insignia in use in 1936.
G. Trade badge for a Troop NCO Saddler (Truppensattlermeister); worn on right forearm.
H. Motor Driver's Badge of Merit in Bronze; worn on left forearm.
I. Shoulder strap for an Oberleutnant from Infantry Regiment 'Grossdeutschland'.
J. Shoulder strap for an Army NCO of the Field Post System.
K. Gebirgsjäger BeVo quality arm badge; worn on upper right

sleeve. A variation on the normal dark green backed badge.
L. Jäger BeVo quality arm badge; worn on upper right sleeve.
M. Specialist badge, BeVo quality, for an artillery gun layer; worn on left forearm.
N, O, P, Q. Trade badges for Motor or armoured mechanic, 2nd class (N), Technical artisan (O), Motor or armoured mechanic, 1st class (P) and Master technical artisan (Q); all badges worn on right forearm.
R. Arm band worn by soldiers acting as stretcher bearers; worn on right upper arm.
S. Cuff title worn by all members of Infantry Regiment 'Grossdeutschland'; worn on right forearm.
24. The use of camouflage clothing was not as widespread in the German Army as it was in the Waffen-SS, but, as the war progressed, certain such items were manufactured and issued to front-line Army troops, including the type of lightweight camouflaged smock with the distinctive lace-up neck opening shown here being worn by a young Company Leader with five years' war service.

25, 26, 27. Three photos showing German troops and their equipment. Many of the items featured in these photos (and in many other photos reproduced in this book) were common to all troops of all branches of service, for example the steel helmet, the gas mask in its cylindrical container, the water bottle and the drinking cup and bread bag. The way these items, and others, were worn or carried differed according to the need of the individual soldier, but without them a soldier could not exist for any length of time in the field.

▼**25**

26▲ 27▼

▲28

▲29 ▼30

28. A 1941 photograph of an officer (left) and his Wachtmeister. Both men have returned from leading a small raiding party on enemy positions and the strain of combat is evident from their faces.

29. A young soldier moves through reeds and long grass somewhere on the Kuban bridgehead, September 1943. Interestingly, the man is wearing a three-compartment Schmeisser ammunition pouch on his waist-belt but is holding a Luger – a strange mixture.

30. Infantry against machine-guns: German troops, making use of shallow cover scraped out of the side of a river bank, move cautiously against a Russian machine-gun emplacement.

31. During the wet season of each year much of the motorized transportation used on the Eastern Front became useless. The summer sun baked the ground hard and with continuous use unmetalled roads were very quickly turned into dust tracks which, with the onset of the wet winter weather, very rapidly became thick, almost impassable mud baths. Heavy vehicles sank into the mud, became stuck fast and had continually to be dug out by hand or towed out by tracked vehicles. All this effort put a strain on manpower and on the engines of the vehicles, the latter thereby consuming far more petrol and diesel fuel than would have been used under normal conditions. Above all, delays in the movement of troops and munitions were caused. The front of this staff car is caked in mud thrown up from the rear wheels of the vehicle used to tow it out.

31 ▼

▲32 ▼33

32. The Schwimmwagen, an amphibious, four-wheel-drive vehicle capable of moving both through water and across land, was often put to good use and driven through muddy conditions. Unfortunately, there were not enough to equip all front-line units.

33. The Kettenkrad, a motor-cycle with a tracked body, was another useful vehicle capable of tackling all but the worst ground conditions.

34. An artillery piece being towed along a frozen mud road somewhere in Russia. Note the rifles fixed to the front surface of the gun shield as well as other items of equipment and clothing, including kindling wood, being carried on the weapon. Just visible behind the left wheel is a sledge, also being towed and also containing equipment. The gunners accompanying their weapon are all wearing the special winter uniform, its snow camouflage side outermost.

35. Sparsely camouflaged with foliage and covered in layers of fine dust, this convoy of troop-carrying artillery vehicles, both wheeled and tracked, grinds along a seemingly endless dirt road somewhere on the Eastern Front. All the troops appear to be wearing the normal Field Service uniform with the Feldmütze.

▲36 ▼37

36. Infantry being transported on self-propelled assault artillery guns, a practice which not only conserved the energy of the troops but allowed them to be brought forward to the combat area both with speed and together as a unit. However, this method did have its disadvantages: if caught in the open by enemy artillery or mortar fire, or if ambushed by machine-gun or rifle fire, the transported Infantry, bunched together as they were, might sustain heavy losses.

37. The crew of a self-propelled assault gun pose for the cameraman. All are wearing the winter uniform and the semi-official fur caps that were frequently seen during the winter months, especially on the Eastern Front.

38. German Panzers advancing in support of Infantry. Whilst the majority of the Infantry troops are crawling forward on their bellies, a small detachment makes use of the protective cover afforded by the armoured bulk of a tank.

39. Mountain Troop personnel dismounting from a German tank.

38▲ 39▼

40. Relaxation from the pressures of combat vigilance was taken whenever the opportunity afforded itself. A simple operation such as having a haircut was not only necessary but was also the chance for a few moments of relaxation.

41. Field cinemas were set up in those areas where direct enemy action was unlikely to take place, and troops were able to enjoy themselves for an hour or two and forget about the war. Here, fully armed troops cheerfully enter the aptly named 'Winter Garden' cinema.

42. Weapon cleaning was an important task, performed whenever the opportunity could be taken: without care and proper maintenance the equipment upon which the men depended for their lives would soon jam and break down. Here, German Jäger troops clean a heavy, tripod-mounted machine-gun.

◀40

41▲   42▼

43. Unlike the British Army, which had a separate branch of service responsible for general catering needs, German troops relied upon cooks who were part of every regiment for the supply and preparation of foodstuffs. Every soldier was issued with a set of mess tins, a water bottle with cup, a bread bag (a form of satchel that amongst other things could hold a small loaf of bread) and a set of eating utensils. Hot food, prepared at mobile field kitchen units known as 'Goulash Cannons', was carried in a special container strapped to the back of the soldier (as in the photograph) and taken to troops guarding outposts in all theatres of war. The men who were charged with this task were specially chosen as they often had to cross exposed ground and visit and move around at least twice a day areas of the front that were under observation and often gunfire from the enemy.

44. Time and military conditions permitting, the German forces made use of bunkers and, if necessary, trench systems. These were usually constructed by combat engineers, and frequently incorporated tree trunks for added strength. Only when the front was stable for a sufficiently long period of time and the need for strongly fortified emplacements arose did the Germans construct concrete fortifications. In this photograph two Grenadiers, crouched in a rough-hewn foxhole, peer anxiously skyward, checking to see if the aircraft overhead is friend or foe.

45. The Commander of an Infantry Division makes a tour of inspection of German emplacements constructed around the city of Leningrad.

46. Making good use of the cover afforded by tree trunks, these Infantry assault troops rest during the fighting in the area south of Lake Ilmen, Russia, May 1942.

▲43 ▼44

47. Wearing his winter uniform, woollen toque and whitewashed steel helmet, a young Grenadier stands guard in a well-stocked outpost complete with rifles and stick hand grenades.

48. Grenadiers manning a well-camouflaged machine-gun dug in alongside a road junction somewhere in Normandy. Note the German armour in the background.

49. Wearing the standard-pattern Field Service uniform and with his basic equipment strapped to his back by means of an 'A' frame, this infantryman crouches in a foxhole, preparing to throw a stick grenade. He is covered by a comrade lying prone and taking aim with his rifle in the background.

50. Wireless or radio transmission as well as field telephone communication was vital for troops operating in the field, and it was particularly important for artillery units. The laying down of field telephone lines under combat conditions was normally carried out by two-man teams. One rifleman carrying a portable cable drum on his back was protected by a second soldier who whilst helping to play out the telephone line also covered his companion in the event of an unexpected attack.

48▲ 49▲ 50▼

▲51

51, 52. Once the telephone line had been laid, it was connected up to the hand sets at both ends and the link between forward units – very often observation posts – and rear areas was possible. Whilst field telephone lines were generally free from interference and were far more secure than radio transmission, they were easily cut, either by accident when run over by a tracked vehicle or deliberately by enemy patrols or during bombardments.

53.

A. Cuff title worn by personnel of Panzer-Grenadier-Division 'Grossdeutschland'; worn on right forearm.

B. Trade badge for a Radio Operator (Funkmeister).

C, D. Various types of badges all worn on right forearm.

E, F. Trade badges for a Fortification sergeant-major; worn on right forearm.

G, H. Trade badges for Motor Transport NCOs; worn on right forearm.

I. Trade badge for a Pigeon-Post master; worn on right forearm.

J, K. Trade badges for Medical personnel (not doctors); worn on right forearm.

L. Shoulder strap for Major of Feldgendarmerie (Military Field Police).

M. Shoulder strap for a Leutnant on the staff of a Mountain Troop school.

N. Shoulder strap for a Leutnant on the staff of Army NCO training school at Sigmaringen.

O. Early-pattern shoulder strap for an Unteroffizier on staff of Nr. 3 Army Group Command.

P. Shoulder strap for a Leutnant from Panzer-Grenadier-Regiment 'Feldherrnhalle'.

Q. Early-pattern shoulder strap for an Unteroffizier from Anti-Tank Regiment 30.

R. Shoulder strap for a soldier from the 'Führer-Begleit-Bataillon', Hitler's personal bodyguard drawn from troops of the 'Grossdeutschland' regiment.

S. Shoulder strap for District Administration Official with the equivalent Army rank of Oberleutnant.

T. Special rank insignia for an Oberfeldwebel. This type of insignia was used on German Army (and Waffen-SS) camouflage and special combat clothing only on the upper left arm.

U. Trade badge thought to be for an Army Administrative NCO (Verwaltungsunteroffizier).

V. Campaign shield issued for the Crimea campaign; worn on upper left arm. Black cloth backing indicates that this shield was worn on the black Panzer uniform.

W. Cuff title for personnel of Infantry Regiment 271 'Feldherrnhalle'; worn on left forearm.

A

B   C   D   E   F

G   H   I   J   K

L   M

N   O

P   Q   R   S   T

U   V

W

Deutsche Wehrmacht A

L 651 B

VI C

201 D

E

W F

H G

W VII H

H I

J

K

L

4 M

Fb N

III O

Feldpost

4.

A. Arm band worn by civilian or paramilitary personnel, German or foreign, employed within the German Armed Forces, including the Army; worn on left upper sleeve. The black-edged green bar is a rank badge for an Unteroffizier of the type usually worn on the German Army camouflage and special combat clothing.

B. Shoulder strap for a soldier training with the 651st Infantry Instruction Regiment.

C. Early-pattern shoulder strap for a soldier on the staff of a recruiting office in Military District (Wehrkreis) VI (Westphalia and the Rhineland; headquarters at Münster).

D. Shoulder strap for a Feldwebel from Infantry Regiment 201.

E. Early style of shoulder strap for a Stabsmusikmeister (Director of Music ranking as a Captain) from Infantry Regiment 73.

F. Shoulder strap for an Oberfeldwebel from Guard Battalion Berlin.

G. Shoulder strap for an Unteroffizier from Infantry Regiment 'Grossdeutschland'.

H. Shoulder strap for an Unteroffizier on the staff of a military recruiting office in Military District VII (Munich).

I. Shoulder strap for a Grenadier from Infantry Division 'Grossdeutschland'.

J. Shoulder strap for a Remount Official with the rank equivalent to an Army Oberleutnant.

K. Shoulder strap for an Army Archivist with the rank equivalent to an Army Hauptmann.

L. Shoulder strap for a Technical Official such as a Building, Engineering or Armoury Official with the rank equivalent to an Army Major.

M. Shoulder strap for an Oberleutnant from Panzer Regiment 4.

N. Shoulder strap for an Ordnance Officer with the rank of Oberleutnant.

O. Shoulder strap for a Major der Landwehr from Military District III (Berlin).

P. Shoulder strap for a Sonderführer (Specialist Officer).

Q. Cuff title for Field Postal Officials; worn on right forearm.

55. Field telephone lines strung overhead at a divisional headquarters to keep them out of the way of men and vehicles. During winter these lines became covered with snow and ice and had to be periodically shaken free of such encrustations to lighten the cables and prevent them from snapping.

◀54 55▼

▲56

◀57

56. A German assault unit from an Infantry and an Engineer Regiment. The officer (back to camera) is accompanied by his runner, who is operating a field telephone, and by a two-man field radio team. In the latter, one man carries the radio strapped to his back whilst the second (wearing earphones) operates the set.

57. Allied troops often had to serve and fight in jungle-covered terrain and were thus liable to be plagued with mosquitoes and other sickness-bearing insects, but the German forces seldom encountered such hazards. However, there were forests and swamplands in southern Russia and deserts in North Africa where, depending on climatic conditions and seasonal weather changes, malaria-bearing mosquitoes, gnats and infection-spreading flies could be a nuisance and quite often a danger to health. German soldiers serving in these areas were thus issued with fine net head-coverings; this photograph shows a Mountain Troop soldier on the Eismeer Front in Russia wearing such a net.

58. The German Military authorities made full use of railway systems. Not only was the German national railway network (Reichsbahn) used extensively for troop, matériel and munitions movements, but the railway systems of annexed and occupied countries were rapidly brought into full use to help in the German war effort. Towards this end, track, rolling stock and buildings had in many instances to be repaired and maintained, and sometimes even specially adapted to suit the needs of the German railway system. Above all, millions of miles of vulnerable railway track and many important railway installations had to be guarded against ground and air attack and against sabotage. Here, German engineers carefully check the lines at a newly captured railway station, 1940.

59. In rear areas, where the threat of air attack was limited and the weather was mild enough, it was possible to move troops around in open freight cars. Here, unarmed troops somewhere in a quiet sector behind the Eastern Front seem to enjoy their move by railway.

58▲ 59▼

▲60 ▼61

60. Heavily protected military trains were necessary to repel attacks by marauding bands of Russian partisans, who not only mined railway track but also attacked and in many instances destroyed vital rolling stock. To counter these attacks, the Germans introduced armoured trains, either trains (i.e. engine and wagons) with additional protection or else specially adapted rolling stock which acted as armoured strongpoints and formed part of a line of carriages or freight cars.

61. The swastika flag was used by military units as a recognition signal. It was tied on to vehicles in a position that could clearly be seen from the air and was also spread out on the ground by infantry units. In this manner German airmen were able to recognize their forward mechanized units and also to distinguish the extent of German positions. The photograph shows German armour in the Donez region. The armoured car on the right displays a swastika flag fastened over the rear engine cowling, a position clearly visible from the air but not easily seen by the enemy from in front.

62. The Germans' failure to occupy warm and protected winter quarters in Russian cities and important towns during the winter of 1941–42 led to their suffering heavy and unacceptable losses, mostly through frostbite and hypothermia. Not only was the German offensive brought to a halt short of its winter target, but the fighting troops were inadequately clothed to face up to the sub-zero temperatures that were commonplace during Russian winters. A vast national drive was instigated in Germany and many of the occupied European countries under the leadership of Dr. Goebbels and controlled by the NSDAP system whereby all forms of warm clothing were collected from civilians volunteering these items: ski clothing, woollens of all types, furs, animal skin coats, top coats, extra thick blankets, silk clothing, in fact anything at all suitable was pressed into service. The troops to which these items were issued took on a slightly odd, un-military appearance, but anything was preferable to freezing to death. Not until the following winter did the essential fighting troops receive purpose-made warm winter clothing. The photograph shows German Grenadiers assisting each other to adjust their equipment. Both men are wearing the reversible winter uniform, introduced for the Russian winter campaign of 1942, with the white, snow camouflage side worn outermost. The special reversible mittens which were part of the issue uniform were held in readiness when not actually being worn by being strung around the wearer's neck on the end of a length of cloth tape, sometimes inside the jacket with the tape threaded through the arms. With almost constant wear, the white material of these winter uniforms very rapidly became dirty.

63. German Infantry and Self-Propelled Artillery soldiers study a map. These troops are wearing specially produced white cotton coveralls, and the soldier pointing at the map with his bare hand is wearing a two-piece snow suit. These garments afforded good snow camouflage and because they were of simple manufacture (unlike the thick, blanket-lined winter uniforms normally worn under these cotton garments) it was an easy matter to have them washed and dried once they became soiled.

62▲

64. Taking cover behind a destroyed T34 Russian tank, German Grenadiers pause for a smoke; all wear the winter uniform with the white side outermost. Clearly seen on the left arm of the soldier in the foreground is the identification arm band buttoned to the sleeve of his white camouflaged winter jacket. These coloured arm bands were introduced by the Germans in an attempt to distinguish between German and Russian troops when both sides were clad in similar white snow camouflage. This visual system operated in much the same way as passwords: the cloth arm bands were repeatedly changed so that different coloured bands were worn at various times, to act as an additional security measure.

65. Top coats and other garments were often manufactured from animal skins with the fur or fleece worn on the inside, although such items were not a universal issue and their use seems to have been favoured by senior officers and certain personnel who could afford them and whose military function would not be impeded when wearing them. Oberst (Colonel) Graf von Strachwitz is shown here wearing a long, fleece-lined coat and talking with a young Panzer-Grenadier officer wearing a jerkin and fleece-lined overtrousers.

66. Soldiers from a battle group about to climb down from an open truck. Some idea of the miserably cold conditions of a Russian winter can be judged from this photograph. Despite the men being clothed in quilted and padded winter uniforms and being huddled together for warmth and protection, the driving snow has compacted itself into every fold and crevice of their clothing and equipment. The chill factor of the freezing air was pushed down many degrees below zero by the driving wind, and cold such as this caused many deaths.

◀64   65▲   66▼

70▶

67. Men from the 'Grossdeutschland' Division, all clad in winter uniforms with the camouflage-patterned side worn outermost, cautiously peer around the corner of a burnt-out shell of a house.

68. Generalleutnant Richter RKT (right), accompanied by his aide, inspects a knocked-out Russian tank and the remains of its crew. Both men are wearing the special winter uniform, and the General is distinguished by the special rank insignia worn on the upper left arm.

69. Gunners from a Mountain Troop Regiment operating their weapon in extreme cold. The soldier on the left is wearing the extra thick, blanket-lined greatcoat with built-in cloth hood, and both men are wearing fleece-lined fur caps, an idea adopted from Red Army troops.

70. A guard posted on a heavy gun somewhere on the Eastern Front. The sentry is wearing the long, very thick, fleece-lined guard coat absolutely necessary for the survival of men who were required to stand in exposed positions for any length of time in sub-zero temperatures.

▲71 ▼72

▼73

1. Troops who were engaged in extended periods of combat rapidly assumed a quite different appearance from that seen during peacetime manoeuvres or even during wartime training exercises. Looking particularly weary, these German non-commissioned officers pause to eat their meagre rations during a move towards the front. All three soldiers are wearing the Model 1938 Field Service cap; the two seated NCOs are wearing their caps with the side curtains pulled down over their ears.

2. An Infantry Feldwebel armed with a Russian sub-machine gun calls his men forward during a particularly bitter period of street fighting for the Russian city of Rostov, August 1942.

3. As the war progressed into its second and third years, the need grew to recognize by visual means the increasing number of acts of bravery and courage displayed by individual soldiers. This brought about a sub-division of some existing military awards as well as the introduction of higher grades of these same awards and of new awards to acknowledge various military acts not hitherto covered. The photograph shows a Company Officer helping to push a Panji wagon out of soft earth. He has been awarded the bronze grade of the Close Combat Clasp (worn above the left breast pocket), the Iron Cross 1st Class, the Infantry Assault badge in silver and the Wound Badge in black (all on the left breast pocket) and the German Cross in Gold (on the right breast pocket). Both the Close Combat Clasp (Nahkampfspange) and the German Cross (Deutsches Kreuz) were examples of awards instituted during the war years, the 'Clasp' being introduced on 25 November 1942 and the 'Cross' on 28 September 1941.

74. General von Seydlitz talking into a field telephone during the march on Naowa. Both the General Officers' and the Officers' version of the Model 1938 'Feldmütze' – the Field Service cap – can clearly be seen: the General's cap is piped in gold and that of the junior officer nearby in silver.

75. Generaloberst Rendulic, holder of the Knights Cross with Oakleaves and Swords, at a command meeting with officers from his Mountain Troop Army command. Rendulic wears the correct version of the General Officers' Field Service uniform, complete with the Generals' peaked cap, whilst the officers with him are wearing the officers' version of the 'Bergmütze' or Mountain cap.

74▲

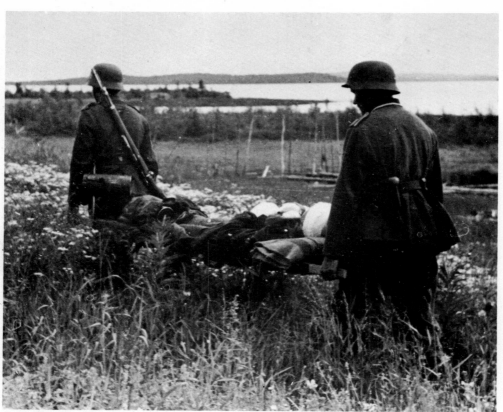

76. The German military medical service performed a very important task. First aid posts, clearing hospital field operating hospital and rear base hospitals all featured in the network of military medical services which stretched back from the front line to the German interior with its military hospitals and convalescent homes. Here, a wounded soldier is carried on a stretcher to a collecting point, prior to evacuation to a field hospital.

77. In difficult terrain such as hills and mountains, where wheeled vehicles could not be used, both mule and horse transport were utilised to bring out the wounded. Shown here is a stretcher case loaded on to a pallet slung between two mules.

▲76 ▼77

78. A wounded Grenadier gets a light from his Battalion Commander. Both men are from Panzer-Grenadier-Division 'Grossdeutsch-land', as can be seen from their cuff titles and shoulder strap insignia. The young Grenadier with his right arm in a splint and sling would be classed as 'walking wounded' but would need to convalesce in order to be able to use his arm again and would thus be away from his unit for a number of weeks.

79. A stretcher case being loaded into a field ambulance.

78▲   79▼

▲80 ▼81

80. The rules and regulations governing the interrogation, welfare and safekeeping of captured enemy soldiers were set out under Appendix 24 of the 'International Convention Relative to the Treatment of Prisoners of War' signed at Geneva on 27 July 1929. However, this 'Geneva Convention', as it came to be known, was not ratified by the USSR, and as a result the vast majority of Red Army troops, including naval and air force personnel, that were taken prisoner by the Axis armed forces (Germany and Italy both signed the Convention) were very badly treated whilst in captivity, many hundreds of thousands of them dying from starvation, unattended sickness, ill treatment, overwork or outright murder. Here, members of the German Field Police guard 240 Soviet prisoners being marched back to a rear area during the initial phase of the attack on Russia.
81. A captured Soviet officer being interrogated by German staff officers from an Infantry Division.
82. A Red Army prisoner receiving treatment from his German captors for a light wound.

82▼

3. A fifty-one-year-old Russian taken prisoner by a German Mountain Troop Regiment undergoes initial interrogation.
4. Horses were employed in large numbers by the German Army, not only as cavalry mounts but also for horse-drawn transport and for towing a whole variety of wheeled vehicles and heavy weapons. On the Eastern Front horses were supplemented by the use of Panji ponies, animals native to Russia and used extensively by Russian peasants and farmers. The photograph shows a horse-drawn ammunition limber passing a group of German Infantry resting on the roadside somewhere in the Leningrad area.
5. Horses pulling wooden sleighs loaded with equipment, February 1942.
6. Panji wagons pulled by horses and containing lightly wounded members of a Mountain Infantry unit.
◀83

84▲

85▲   86▼

▲87 ▼88

89▲

90▲

87. Mules were another essential form of animal transport. Being capable of climbing and crossing areas where wheeled or tracked vehicles were unable to venture, these animals were ideally suited to the needs of Mountain Troop units.

88. Wearing white snow smocks over their winter greatcoats, men from a Rifle Division (Jägerdivision) lead a column of pack mules loaded with weapons and equipment along a road near Stanislau.

89. In an effort to bring down a Soviet Air Force reconnaissance plane this machine-gun was taken into improvised use as an anti-aircraft weapon. Resting the weapon on the shoulder of the NCO (who also holds firm the tripod legs) the gunner takes aim and fires upwards in this position. This practice had a bad effect on the soldiers' eardrums: prolonged or excessive noise such as gunfire or violent explosions very often caused severe damage to the hearing of troops and sometimes permanent deafness, but the German forces made no universal effort to prevent such impairment.

90. A heavy mortar being fired from a snow-covered, shallow-dug firing pit, April 1942.

▲91

▲92

▲93   94▶

91. Wearing full battle order, German Infantrymen move inland from the Kuban bridgehead, October 1942.
92. His windproof anorak with the hood turned up, hand grenades stuffed into his waist-belt, and rifle lying by his side, this young 'Ostfront' soldier snatches at sleep during a lull in the fighting.
93. Working in their shirtsleeves, German 'Smoke Troops' load up a multi-barrelled mortar.
94. A heavy machine-gun mounted on its anti-aircraft tripod is manned by a sentry who scans the skies through his field glasses.

95. Wearing a privately procured sheepskin overcoat, Oberst Graf von Strachwitz talks to a combat Pioneer soldier equipped with a flame-thrower.

96. A patrol leader somewhere on the Italian Front armed with an anti-tank mine.

97. An anti-tank gun and its crew in action at Stalingrad.

98. A lucky miss. This NCO examines the remains of his Luger holster, the top cover of which has been torn away by a splinter from a Russian mortar.

99. German Grenadiers wearing camouflage jackets and cloth covers to their steel helmets examine the interior of a British Army armoured car knocked out during the Normandy fighting. The vehicle was from the 51st Highland Division, as can be seen from the 'HD' formation sign painted on the left-hand front wing.

100. Dogs were frequently used as message-carriers by the German ground forces, the message being contained in a small metal cylinder attached to the dog's collar (it can be seen in the photograph under the animal's throat). Also strapped to a small harness is a mess tin and a water bottle. Note the dog's muzzle carried on the waist-belt of the dog handler.

▲95 ▼96   ▼97

98▲

99▲   100▼

▲101 ▼102

103▲

101. A German Infantry Assault Unit (Stosstruppe) carefully works its way across open ground towards a row of small trees. All the men are wearing Field Service uniforms and carry the minimum amount of necessary equipment.

102. A German machine-gun and observation post somewhere on the front line surrounding Tobruk. The troops are wearing the original DAK sun helmet which was eventually replaced by the more practical and popular peaked Field cap, an example of which can just be seen on the soldier in the background.

103. After the armoured thrust comes the Infantry: German machine-gunners cross frozen ground at speed, February 1942.

104. A German sniper somewhere in the Crimea. This officer (note the shoulder strap, just visible) is wearing a set of 'D' ring 'Y' straps over a hard-wearing version of the Officers' Field Service tunic, which has the unusual added feature of a two-button, cloth half-belt at the back.

◄104

▲105

105. German troops of the Afrika Korps are inspected by Italian General Gariboldi and their Commander General Rommel, April 1941. The men are parading in the type of uniform worn during the early days of the Deutsche Afrika Korps. Although this was designed as a tropical uniform it proved to be unsuitable for desert warfare: the olive green colour of the jacket and breeches was wrong for the desert background; the garments were too formal and restricting; the lace-up boots were too complicated to wear; and the sun hat proved very unpopular. All these items were eventually replaced by other, more suitable clothing.

106. German troops were always accompanied by Field Police: an NCO from a Feldgendarmerie unit here mans a roadside checkpoint on the route to El Alamein.

107. The 'Feldgendarmerie' operated in all theatres and in all countries. The photograph shows a three-man unit, all NCOs, putting up signs somewhere in France. When on duty, troops of the Field Police wore a gorget suspended from a chain from around their necks.

▼106

107▶

108. In an effort to produce a form of clothing for crews of armoured vehicles (including tanks) that was cheaper than the existing black or field-grey short cross-over jacket and had better camouflage properties than the Black Panzer uniform, a reed-green denim jacket was manufactured. This had the distinctive addition of a large patch pocket on the left-hand side of the chest,

a feature which can clearly be seen in this photograph.
109. A motor-cycle despatch rider delivers a sealed message to a Company officer. The rider is wearing the special rubberized motor-cycle coat, together with motor-cycle overmittens – the standard-issue garments worn as protective clothing by these personnel.

▲110

110. Helmet covers produced
from camouflage-printed cloth
became a common feature in
the German Army during the
war. They were manufactured
from panels of cloth sewn
together to follow the contours
of the steel helmet and were
held in place by being gathered
in around the edge of the
helmet and tied up under the
rim. Cloth loops were also
attached to the covers to enable
the helmet to be garnished
with foliage and grass.

111. A disabled Tiger tank
undergoing field repairs:
mechanics and the crew of the
tank work to replace a damaged
track. The tank is supported
over an improvised work-pit
by a heavy duty jack raised on
blocks of timber. Note the
section of track and the
smashed gun barrel in the
foreground. The men working
on the repairs have a mixture
of uniform items, some
wearing all or part of the
special black tank suit and
others wearing field-grey or
rush-green denim items.

112. French farmers and
farm-hands look on with
impassive interest as German
armour moves into southern
France, November 1942.

▲113

113. An anti-aircraft post somewhere in the Crimea. The sentry posted as look-out scans the sky through field glasses, searching for possible approaching enemy aircraft. The markings painted on the gun shield testify to the skill of the gun crew in having shot down fifteen enemy aircraft.

114. Almost from the beginning of the June 1944 Normandy

landings, the Allies enjoyed air superiority. This meant that nothing the Germans attempted to move, be it men, armour, gun or transport, was safe from destruction by aerial gunfire, bombin and rocket strafing. In an effort to conceal their presence or movement and thus avoid attack, the Germans made much use of natural camouflage.

▼114